Old Buckingham Station
Chesterfield, Virginia

Investigated by: Thomas H. Miller, P.E.

This is Report 105 of the Major Fires Investigation Project conducted by Varley-Campbell and Associates, Inc. under contract EMW-94-C-4423 to the United States Fire Administration, Federal Emergency Management Agency.

Department of Homeland Security
United States Fire Administration
National Fire Data Center

U.S. Fire Administration Fire Investigations Program

The U.S. Fire Administration develops reports on selected major fires throughout the country. The fires usually involve multiple deaths or a large loss of property. But the primary criterion for deciding to do a report is whether it will result in significant "lessons learned." In some cases these lessons bring to light new knowledge about fire--the effect of building construction or contents, human behavior in fire, etc. In other cases, the lessons are not new but are serious enough to highlight once again, with yet another fire tragedy report. In some cases, special reports are developed to discuss events, drills, or new technologies which are of interest to the fire service.

The reports are sent to fire magazines and are distributed at National and Regional fire meetings. The International Association of Fire Chiefs assists the USFA in disseminating the findings throughout the fire service. On a continuing basis the reports are available on request from the USFA; announcements of their availability are published widely in fire journals and newsletters.

This body of work provides detailed information on the nature of the fire problem for policymakers who must decide on allocations of resources between fire and other pressing problems, and within the fire service to improve codes and code enforcement, training, public fire education, building technology, and other related areas.

The Fire Administration, which has no regulatory authority, sends an experienced fire investigator into a community after a major incident only after having conferred with the local fire authorities to insure that the assistance and presence of the USFA would be supportive and would in no way interfere with any review of the incident they are themselves conducting. The intent is not to arrive during the event or even immediately after, but rather after the dust settles, so that a complete and objective review of all the important aspects of the incident can be made. Local authorities review the USFA's report while it is in draft. The USFA investigator or team is available to local authorities should they wish to request technical assistance for their own investigation.

This report and its recommendations were developed by USFA staff and by Varley-Campbell & Associates, Inc. Miami and Chicago, its staff and consultants, who are under contract to assist the Fire Administration in carrying out the Fire Reports Program.

The U.S. Fire Administration greatly appreciates the cooperation and information received from Chief Robert L. Eanes, C.E.M. and Battalion Chief James E. Graham, Chesterfield Fire Department, and William D. Dupler, County of Chesterfield, Virginia.

For additional copies of this report write to the U.S. Fire Administration, 16825 South Seton Avenue, Emmitsburg, Maryland 21727. The report is available on the Administration's Web site at http://www.usfa.dhs.gov/

U.S. Fire Administration
Mission Statement

As an entity of the Department of Homeland Security, the mission of the USFA is to reduce life and economic losses due to fire and related emergencies, through leadership, advocacy, coordination, and support. We serve the Nation independently, in coordination with other Federal agencies, and in partnership with fire protection and emergency service communities. With a commitment to excellence, we provide public education, training, technology, and data initiatives.

 FEMA

TABLE OF CONTENTS

Old Buckingham Station
58 Unit Apartment Building
Chesterfield, Virginia
May 19, 1995

Investigated by: Thomas H. Miller, P.E.

Local Contacts: James E. Graham, Battalion Chief
                   Chesterfield Fire Department
                   Fire Department Headquarters
                   P. O. Box 40
                   Chesterfield, VA 23832

                   William D. Dupler, Building Official
                   County of Chesterfield
                   Department of Building Inspections
                   P. O. Box 40
                   Chesterfield, VA 23832

OVERVIEW

On Friday, May 19, 1995 at 1:46 a.m., the Chesterfield County Emergency Communications Center received a 9-1-1 telephone report of a fire in the sprinklered Lodge Building at the Old Buckingham Station apartment complex. This X-shaped, three and four story, wood frame structure was the largest building in the complex and contained the management offices, social function room, and 58 apartments.

All residential buildings in the complex were protected by an automatic sprinkler system. The system design basis was a modified version of NFPA Standard No. 13D, and the Standard for Sprinkler Systems in One-and Two-Family Dwellings. The 13D Standard was not intended for use in multifamily dwellings and NFPA 13R, which applies to multifamily dwellings up to and including four stories high, was first published in February 1989, after the complex was constructed. Details on the automatic sprinkler protection are provided later in this report. The sprinkler systems were connected to the domestic water service and through a 1-1/2-inch water meter, as would have been appropriate, not a low friction loss fire type meter. Some water meters have significant pressure losses and the NFPA sprinkler installation standards indicate that these losses must be included on the design calculations. Special fire meters are used to minimize pressure losses, especially at high water flow rates.

The first company on the scene found fire through the roof of the building's center section with the fire spreading into all four building wings simultaneously. Fire department resources were divided between fire suppression and occupant search and rescue. Complicating the suppression effort was the location of the fire's greatest involvement on the side of the building without direct street access, making rapid placement of hose streams directly on the fire more difficult.

By the time the fire was declared under control at 4:53 a.m., more than 3 hours later, over 30 apartments were completely destroyed, another 12 sustained heavy damage and the balance suffered smoke and water damage. None of the apartments could be occupied after the fire. The Chesterfield Fire Department committed 13 engines, 2 trucks and 5 special service units staffed by 76 officers and firefighters to this incident. Estimates of the fire loss are 4.4 million dollars for the building and 1.1 million dollars for tenant property.

KEY ISSUES

Issues	Comments
Large Unsprinklered Combustible Spaces	The lack of automatic sprinklers in substantial combustible spaces allowed a large fire to develop. The 1996 edition of NFPA 13R permits unsprinklered open balconies, stairs and corridors, outside porches, attics and concealed spaces not intended for living purposes or storage
Inappropriate Sprinkler Standard Used	NFPA 13D is not intended for large structures or for other than use in 1 and 2 family dwellings and mobile homes.
Draft Stopping and Tenant Fire Separations	The attic represented a substantial unprotected combustible space whose primary fire defense is the building code required draft stopping at tenant separation walls. Between effective draft stops, unimpeded fire growth can be expected and, with typical attic ventilation, plenty of air will be available for this growth. Stopping the fire's spread at this point will depend on the integrity of the draft stops, fire department intervention, and the failure time of the construction. In this structure, the location of draft stops may not have coincided with tenant separation walls in all locations. The fire may have traveled through apartments under the draft stops.
Ventilation Openings into Attic	Air vent openings at the building eaves permitted the fire to easily access the unprotected combustible attic space.
Fire Spread	The open combustible corridors and balconies and lightweight exterior wall finish supported rapid vertical and horizontal fire spread.
Delayed Fire Reporting	The exterior point of origin and time of ignition combined to cause a long delay in the discovery and reporting of this fire.
Access Limitation	The main body of the fire was on the side of the structure with restricted access. Swimming pool, fencing, and lack of roadways delayed the fire department's attack on this side.

BUILDING HISTORY AND CONSTRUCTION

The Old Buckingham Station apartment complex consisted of three and four story wood frame multi-family structures distributed over a wooded site with rolling hills. Buildings are grouped in clusters at different elevations. Vehicle access within the complex is provided by paved roadways and parking areas adjacent to the buildings. Two connection points provide access to the Midlothian Turnpike which passes on the south side of the complex. (See Appendix A for site plan.)

According to Chesterfield County records, preliminary pre-design meetings with the developer and architect began in September 1986. While automatic sprinklers were not required by the building code, modified residential automatic sprinkler protection for the entire complex was discussed from the outset. These discussions related to the building code "modifications" that would be allowed based on the added protection provided by the sprinkler installation. (A description of the modifications granted is provided in the Building Code Section of this report.) Construction began in June 1987 and a certificate of occupancy was issued in January 1989. All of the buildings were protected by what local officials describe as a modified NFPA 13D residential style automatic sprinkler systems.

Fire Building

The Lodge Building was the largest building in the complex and contained 58 apartment units with over 57,000 square feet of occupied floor area. The building varied from three to four stories in height. The building was "x" shaped with a four story center core area and four, three-story wings. (See Appendix B for drawing.) The core contained the complex's management offices and social function room on the lower two floors and two apartments on each of the third and fourth floors. A single elevator in the core area served all four floors by means of open corridors.

Two of the three-story wings had approximately 4,700 square feet of area on each floor divided into five apartments. These larger wings formed the east front of the building. The two rear wings were also three stories high but contained only four apartments and about 3,500 square feet per floor. The floor-to-floor height was approximately 9.5 feet. A common, multiple level peaked roof with asphalt shingles covered the core area and the four wings.

The exterior walls were load bearing wood stud construction covered with vinyl siding over Thermo-ply sheathing, a thing specialty construction material which has the visual appearance of pressed paper board covered with aluminum foil. These walls were insulated with fiberglass batts in the studs spaces, with a sheet plastic vapor barrier covered by gypsum board on the inside. The interior walls were also of wood stud construction covered with gypsum board.

The floors were 2 x 4 parallel cord trusses 18 to 24 inches deep supporting a plywood sub-floor that was topped with a thin (about 1-inch) layer of lightweight concrete. Gypsum board ceilings were attached directly to the bottom cords of the truss. The roof was constructed of a chip board deck over peaked wood trusses made from 2 x 4 and 2 x 6 members. The truss span was approximately 40 feet with a peak height of 13.5 feet and an 8/12 pitch. The roof contained a ridge vent and continuous soffit vents were provided around the perimeter of the building. (See Appendix C for soffit vent details.)

The combustible attic space was divided by draft stops at intervals of every three or four apartment units depending on the wing of the building. The draft stop construction was indicated as a single layer of 1/2-inch-thick gypsum board attached to one side of the roof truss. The floor trusses were draft stopped at the separation walls between apartments. This was reportedly done by continuing the gypsum board wall covering to the floor deck above rather than stopping at the ceiling attached to the bottom truss cord.

Each apartment entrance door opened directly to the outside. The first floor apartments were accessed via a covered concrete sidewalk. The upper floors had wooden covered walkways leading in two directions to open wood frame stairs one located at the core and the other at the ends of the wings (See Appendix C for diagram). The walkways were of wood construction and contained small gaps between the deck boards to permit water to drain through. With the exception of horizontal side-wall automatic sprinkler heads that were located near the two core stairways, the covered walkways were not protected by automatic sprinklers.

On the side of the apartment wings opposite the walkways, each unit had a covered exterior balcony. On the first floor, the balconies were concrete on grade, while the upper floor balconies were wood frame construction very similar to the walkways. There was no automatic sprinkler protection for any of the balconies.

BUILDING FIRE PROTECTION

Automatic sprinklers were installed in all of the apartments, storage, and other occupied spaces. The installation in the apartments followed a modified NFPA Standard No. 13D wet pipe automatic sprinkler design. This system was designed and installed before NFPA Standard No. 13R was adopted. (See Appendix D for an explanation of NFPA 13, 13D, and 13R.) The residential sprinkler systems were supplied by a 1-1/2-inch metered domestic water connection into each of the four wings. Each wing had a separate, single 1-1/2-inch hose connection for the fire department to supplement the sprinkler water supply. These connections were located in the first floor walkway which surrounded the core area. There were no standpipes or hose valves connected to any of the automatic sprinkler systems.

The sprinkler system used residential, quick-response sprinkler heads supplied by polybutylene piping in the apartments. The installation employed Underwriters Laboratories listed products and connection methods utilizing heat fusion rather than the mechanical type connections used with polybutylene plumbing.

An NFPA Standard No. 13 compliant wet pipe automatic sprinkler system was installed in the core offices and social function room. Residential quick response heads were used in the third and fourth floor apartments; regular sprinkler heads supplied by steel piping were installed in the offices and social room. This system had a separate code-complying fire protection water supply connection with a fire department connection located at a pit in front of the building. This water supply was also connected to the residential sprinklers in the four apartments on the third and fourth floors of the core section.

The combustible concealed spaces, including the attic, floor/ceiling space and vertical shafts were not protected by automatic sprinklers. The outside wooden walkways and balconies were also unprotected, except for some selected core areas. At these core areas, the two open stairways were protected by dry-type sidewall automatic sprinkler heads located on the wall opposite the stairways. Evidence indicates that one of these sprinklers operated to extinguish the fire at the point of origin. However, by the time the sprinkler operated, the fire had spread beyond the sprinkler head's reach.

Each apartment was also provided with a hardwired single station smoke detector located outside the sleeping areas. Alarm bells for occupant notification were operated by water flow switches in the automatic sprinkler system. Manual pull stations were not provided.

Automatic Sprinkler Water Supply

The automatic sprinkler systems and fire hydrants in the Old Buckingham Station complex were supplied by an 8-inch ductile iron water main loop that was connected to a 12-inch circulating water main under Midlothian Turnpike. Because of the rolling type hills in the county, water system pressure varies significantly. Water flow test information from 1988 indicates that 1800 gpm was available at 20 psi in the area of the complex. Static water pressure for this test was recorded at 58 psi for a high elevation. Adjusted to the Lodge Building elevation, this pressure would have been about 68 psi. The modified sprinkler system was tested after installation by flowing water from four remote heads. This test indicated that the supply met or exceeded NFPA requirements. (See Appendix D for requirements.)

THE FIRE

Fire investigators determined that the fire began on the second floor in a covered walkway between the northeast wing and the center core, just outside of apartment 2J. (See Appendix B for location.) The cause was identified as an exterior light fixture that had been inverted from its usual operating position by one of the occupants and then wrapped with a flannel shirt to reduce the light intensity into their bedroom. The fixture's incandescent bulb was enclosed by a clear plastic globe. By inverting the fixture, the vent holes, which were intended to remove heat from the 60-watt bulb, were out of position and the plastic globe in combination with the flannel shirt held in the heat. This eventually resulted in the ignition of the shirt and the light fixture's plastic globe. The fire spread to the vinyl plastic siding and to the wooden stairway.

The siding fire quickly moved up the nearby open, wooden stairway before the sidewall sprinkler on the wall opposite the light fixture was able to operate. (See appendix E for Section Drawing.) The fire continued to burn upward until it reached the ceiling of the fourth floor open walkway. At this point it entered the combustible attic through a vent opening at the outside perimeter of the ceiling. The soffit vents in the fourth floor core area were located inside, rather than outside, the perimeter support beam for the roof. This design contained the heat and flames at the fourth floor ceiling and allowed them to pass through the vent into the unprotected combustible attic. Once in the attic, the fire was out of reach of the exterior sprinkler heads.

As the fire burned in the unsprinklered attic, it was able to quickly travel horizontally to the fire/draft stops and then vertically up through the roof deck and down into the apartments.

The fire consumed the structural supports for the roof, ceiling, and automatic sprinkler system. As these items failed, the fire moved past the fire/draft stops into the next attic section. The fire travel paths included burning across the combustible roof as fire/draft stops do not penetrate the roof deck. In addition, the fire may have also burned around the bottom of the draft stops as some of these stops may not have been placed in line with the separation walls between apartments.

The fire also spread horizontally through the unprotected open wooden walkways around the building's core. (See Appendix D for a comparison of automatic sprinkler requirements.) This allowed the fire to spread to all four wings and all four floors through the core's open stairways. The fire also began dropping down from the attic to involve the unprotected combustible balconies, which allowed the fire to spread into multiple apartments through the large glass doors which opened onto the balconies. By the time this fire encountered automatic sprinklers inside the apartment living areas, it had considerable momentum and was already attacking the sprinkler supporting elements.

Fire Department Rescue and Suppression Activities

Chesterfield's 9-1-1 Emergency Center received a telephone report of the fire and, at 0147 hours, dispatched an initial assignment from Stations 4 and 5. Two engines from each station, Trucks 37 and 77 and the North Battalion Chief responded. Initial station alerts also provide firefighter, officer and equipment resources staffed by responding volunteer members. Station 5, about 0.8 mile from the apartment complex, was the closest station dispatched. Engine 53 with five firefighters was the first unit on the scene at 0153 hours, followed by Engine 54 with four firefighters.

On arrival, the officer of Engine 53 reported fire showing from the roof over at least two third floor apartments adjacent to the core area and spreading rapidly horizontally and vertically. Incident Command was established and sectors identified using their standard procedure. The front (east) side of the building was designated as Sector A. The designations then proceed clockwise around the building: Sector B on the south, C on the west and D on the north. Each sector contained one side of two different wings plus part of the core area. (See Appendix F for drawing.)

Incident Command passed from Engine 53's officer (career) to the Assistant District Chief (volunteer) at 0156 hours and then to the area's Battalion Chief, Battalion 3, (career) at 0200 hours. At the change in commands, Engine 53's officer took charge of the Interior Sector. The initial strategic plan begun by Engine 53's officer to search for and evacuate occupants and control fire spread was maintained throughout the incident.

Because of the time of night, it was expected that most residents were home and asleep. Notification and evacuation of the occupants became the first priority of arriving companies. Crews of the initial units were split into two teams; about half of each crew was assigned to search and evacuation and the balance of each crew to fire suppression.

The suppression effort of the first two engine companies, Engines 53 and 54, was to support the core section automatic sprinkler system and to operate a deluge set on Sector A (See Appendix F for drawing.) The core sprinkler system was charged through the siamese connection located at the pit in front of the building. There is no indication that any of the four wing fire department connections to the sprinkler systems were used.

At 0156 hours, the Incident Commander reported "water on the fire" and directed incoming units to prepare for exterior master stream operations and to assist with search efforts. The first truck company, Truck 37, was positioned on the north side of Sector A to operate an elevated master stream. Engine 43 was directed to Sector D and to place into operation a portable deluge set at this sector's west end. (See Appendix G for drawing.) Engine 44 was directed to assist Engine 43 with the deluge; both crews were split into search and suppression elements. As additional Station 5 volunteers arrived, they were primarily assigned to the search and evacuation of occupants. Truck 77 was positioned at the east end of Sector D for the second elevated master stream. At this time, three master streams were operating and a fourth was being set up. Search and evacuation operations were underway in Sectors A and D.

Around 0200 hours, the Incident Commander requested three additional engine companies to the scene. Other engine and truck companies were also being relocated from county fire stations away from the fire to cover nearby stations. These relocations were coordinated by the third on-duty battalion chief whose district was furthest from the fire. Special equipment such as the communications/command vehicle, air mask unit, and lighting were also dispatched during this time.

At 0207 hours, the Interior Sector reported that the fire in the central core area had dropped from the third floor to the second floor and about 4 minutes later, the crew in Sector C reported fire now on all four floors and through the roof at the core. Search of this section was deemed impossible due to the intensely burning fire in this area and the beginning of structural collapse.

Engine 93 was assigned to Sector D to lay a line from Truck 77 to the area of Engine 43 and to supply the line. Engine 102 was also assigned to this sector and connected into the two supply lines to the deluge set at the west end. These parallel lines exceeded 400 feet in length and Engine 102 relay pumped into the deluge. The crews from 93 and 102 were initially assigned to completing the search operation and then later to suppression operations in Sectors B and C.

Chesterfield Fire Department does not routinely use large diameter hose (LDH) for water supply on its apparatus although it has several engines so equipped on an experimental basis. No LDH was used during this incident. The department's usual water supply hose on engine companies is 3 inch diameter double jacket lined hose. Unless specially noted, supply lines from hydrants to the fireground were 3 inch diameter.

Engine 53 was moved from Sector A in front of the building to Sector B. It laid parallel supply lines from Engine 54 to its position in Sector B. This crew and a number of Station 5 volunteers operated a handline on this side of the fire in addition to searching the apartments for occupants.

Between 0200 hours and about 0225 hours, six engines, two trucks and the Station 5 volunteers who had responded directly to the scene were operating four master streams, two handlines and were searching all of the building not presently involved.

The "All Clear' signal for the primary search for Sector B was given at 0225 hours and for Sector D at 0232 hours. As the principal means of access to the apartments in all four wings were located in these two Sectors, this was also the all clear for the building. (See Appendix B.)

By about 0230 hours, the fire involved approximately 75 percent of the building's entire roof, all of the fourth floor in the center core, about 60 percent of the building's third floor and all of the first and second floors of the center core. Battalion B, Senior Battalion Chief, assumed Incident Command at 0240 hours after touring around the building. Battalion 3 was designated as Operations Officer and the Assistant District Chief was designated the Incident Safety Officer. With the primary search completed, the Incident commander directed all available resources to fire containment and suppression.

At this time effort was made to increase the water supply being delivered to the fireground. The two hydrants nearest the building were being fully utilized by apparatus and the next available hydrants were each over 700 feet away. Engines 103 and 106 combined together to lay a new supply line from Sector A to the hydrant at Buckingham Station Drive and East Coal Hopper Lane (See Appendix H). Engine 103 pumped from the hydrant to Engine 106 which supplied a line into Sector D. The crews initially operated lines on the second floor of this Sector but this operation was later suspended because of the elevated master streams working on both sides of the building's wing in this area.

Engine 102 moved the portable deluge from Sector D into Sector C by extending the lines. The crew stretched a supply line from the engine into Sector C where it was wyed into two 1-3/4-inch hand lines. These handlines were directed into this wing and also used to protect the building at 13201 Boggie Road West which was being affected by radiant heat.

Engine 112 laid a second supply line from Engine 103 to Engine 54 and then into Sector B. Engine 112 directed used its mounted deluge, a 2-1/2-inch handline, and a second master stream onto the fire in Sector B.

Engine 73 was also assigned to increase the water supply to Sector B. Two supply lines were stretched from a hydrant at the rear of the Village Shopping Center to Engine 53. The placement of these lines involved a 600 foot long hand stretch through the woods. Once the lines were in place, the crew assisted with the placement and operation of a deluge from the southwest corner of Sector C into the center of the building.

By 0330 hours, ten engines were supplying water to two elevated master streams, seven additional master streams and six handlines of assorted diameters. By counting the master stream appliances and the number of handlines, fire officials estimated a peak fire flow of 5,500 gpm was being delivered to the fire. The spread of the fire was halted and the heavily involved areas began to darken down. The master streams continued to operate for the next 1-1/2 hours. During this time, crews were also performing secondary searches of apartments. Special emphasis was placed on those apartments where there was no account of the occupants' location.

A severe thunderstorm moved through the area after 0400 hours. The elevated master streams were shut down and the aerial apparatus lowered. Sector Officers were ordered to shelter all personnel in safe locations. The portable deluge sets and engine mounted deluges continued to operate unmanned during the storm. During the fire suppression effort, the Incident commander twice requested all Sector Officers to report on Personnel Accountability. The first was at 0340 hours during the major assault on the fire. The second occurred at 0450 hours after the severe thunderstorm had subsided and crews were returning to the fireground.

At 0453 hours the fire was declared under control. The Operations Officer met with the Sector Officers to develop an overhaul plan. It was decided that only exterior suppression operations would be permitted until day break, when a full safety assessment of the building could be made. The Chesterfield County Building Officials Office was contacted to assist with the building's structural assessment. Sections of the building were identified where it was considered structurally safe to overhaul.

Release of companies began at 0800 hours and a shift change for on scene companies was conducted between 0800 and 0845 hours. Overhaul and salvage operations continued all morning and into the evening. Five engines and two trucks remained on the scene for most of the day.

An Occupant Services Sector under the command of a department captain was established by the Incident Commander during the morning. It was responsible for the coordination of resident access into their apartment to salvage and recover personal items from sections of the building identified as being safe. Residents were escorted into their units starting at about 1300 hours and access continued into the evening hours.

Injuries

The fire department reports indicate that there were no fatalities or injuries to apartment occupants. Two firefighters became ill and were transported to the hospital for observation; they were later released. There were no injuries to firefighters.

BUILDING CODES

At the time of design and construction of the Old Buckingham Station apartments, Chesterfield County was using the 1984 editions of the BOCA Basic National Building and Mechanical Codes. These documents did not require automatic sprinklers throughout any of the buildings planned for the complex. Negotiations between the developer and Chesterfield County officials produced an agreement requiring a modified NFPA Standard No. 13D automatic sprinkler system to be installed in all buildings. This agreement provided for the use of 1/2-inch type "C" gypsum board in place of 5/8-inch type "X" board except at common party walls and fireplace chases. The agreement also allowed that, subject to building separation distances, the exterior walls did not need to have a fire resistance rating and fire rated floor/ceiling assemblies would not be required, allowing the use of 1/2-inch dry wall. Allowable setback distances from streets and parking areas were also increased.

Fire stop partitions in an attic are important to slow the fire's spread throughout this large unsprinklered combustible concealed space. Their purpose in the building code is to provide a brief, rarely more than 20 minutes, period to hold the fire from spreading horizontally beyond the first compartment. Their successful function anticipates that the fire department can be in place and attacking the fire within this time period. Even when constructed in accordance with building code requirements and well maintained (i.e., with no openings or holes around penetrations), they require timely fire department intervention to be effective. Firefighters with hoselines and tools will need to be dispatched to these locations early on the fire. These partitions also have to be located where the fire cannot easily travel around or under them. In it good practice to locate them in conjunction with the tenant separation walls as they are also required to have some resistance to fire spread.

The developer requested that the attic firestops be provided on the basis of 3,000 square feet segments. In the requested attic firestop arrangement, the walls in the attic may not have been located at the same place as the tenant separation walls. According to April 28, 1987 correspondence, the county rejected the request. The tenant separation walls were to continue through the attic space to the underside of the combustible roof deck. Based upon the fire department's investigation after the fire, the attic firestop walls were not in alignment with the tenant separation walls.

FIRE DEPARTMENT AND EMERGENCY SERVICES

The Chesterfield Fire Department is a combination career and volunteer department that operates from 15 stations. It is staffed by 325 career personnel and 350 volunteers. Chesterfield County's population is 233,000 and it measures approximately 446 square miles. In 1994, the fire department reported a total of 17,653 EMS, fire and other emergency calls.

Out of the 15 stations, the department operates some 23 engines, 5 trucks, 4 tankers, 3 heavy duty rescue squads, an airport crash rescue vehicle and a number of specialty vehicles for Hazmat, underwater dive rescue, brush, lighting and air mask support. A combined fire and police mobile command vehicle is also available. The department continuously staffs four advanced life support (ALS) ambulances with career personnel. An additional five ambulances are staffed during the daytime with career personnel and at night with volunteer members. These five ambulances are typically ALS in the daytime and basic life support (BLS) at night. An additional four all volunteer rescue squads are provided in the county.

Career personnel operate in a three platoon system of 24 hours on duty with 48 hours off. The county is geographically divided into three battalions with a career battalion chief on duty in each area. The senior battalion chief on duty is the overall shift commander and typically responds to the scene of any working fire in the county. The senior battalion chief in each platoon reports to the deputy chief of emergency operations. The operations deputy also supervises the emergency medical and med-flight functions. A second deputy chief supervises support services including training and safety, fire prevention, maintenance and logistics, administrative services, and information services. The two deputy chiefs report to the chief of the department who is also the county's coordinator of emergency services.

Minimum staffing on both engines and trucks is three and many companies will typically have four personnel. Staffing is typically higher at nights when volunteers will sleep in the stations along with career staff. Two stations are staffed entirely with volunteers, five stations are staffed entirely with career personnel and eight stations are combination career and volunteer. In these combined stations, typically two separate engines are provided; one for the career staff and the other for volunteers.

Fire hydrant location and spacing will depend upon the area of the county. Although water mains may be present, fire hydrants are installed only as the land is developed along the roadways. The location and spacing would be based on the size and occupancy of the development. Where new water mains are extended into developing land, hydrant location and spacing is usually reviewed with the fire prevention bureau. In this project, hydrant spacing adjustments were made to recognize that all of the apartment buildings were being protected by automatic sprinklers. As a result, the hydrant spacing in the complex exceeded the typical 300 to 500 feet spacing used for multifamily residential buildings.

LESSONS LEARNED

1. **Large unsprinklered spaces with exposed combustible construction can allow a fire to develop and spread beyond the control of automatic sprinklers in adjacent spaces.**

    NFPA Standard Nos. 13D and 13R allow unsprinklered areas with exposed combustible construction such as attics. Once a fire enters such a space, it can spread beyond the reach of the automatic sprinklers protecting the living areas. It was recognized during the development of these standards that some residential fires, once they penetrate into an unprotected space, would not be controlled by automatic sprinkler systems complying with the standards.

    In this incident, the fire started and grew to a significant size before the automatic sprinkler in the area of origin operated. The fire entered the unsprinklered combustible attic space and spread horizontally before dropping into the upper floor apartments. At this time, the fire size was beyond the control of the installed automatic sprinklers. It also spread on the exterior of the structure by means of the combustible balconies, walkways and lightweight combustible siding. This avenue of travel allowed the fire to quickly enter the lower floor apartments through the glass balcony doors and exterior windows.

2. **The building construction used light weight combustible elements which allowed the fire to rapidly spread vertically and horizontally. The unprotected ventilation openings provided an avenue for the fire to readily penetrate into the unsprinklered combustible attic spaces.**

    The fire quickly spread upwards on the exterior of the building, taking advantage of the lightweight plastic siding, an open wooden stairway and wooden balconies. The under-eave vents provided an easy path by which the fire and hot gases could enter the attic. Although the sidewall sprinkler head on the wall opposite the point or origin eventually operated and controlled the fire at that point, the fire had already spread beyond the range of this head. The combustible construction materials and their arrangement allowed the fire to spread faster than the automatic sprinkler heads could respond. (Thermal lag of the sprinkler heads causes an operational delay, consisting of the time between heat first reaching the head until when the fusible element opens.) Location of vents should be coordinated with openings in exterior walls and balconies to minimize the potential for fire spread. Attic eave vents should not be located near windows, doorways or vents which may allow fire extension to the attic.

In addition to the automatic sprinklers, compartmentation is another means of controlling the speed and range of fire spread. To be effective, compartment construction materials should have sufficient resistance to fire penetration and openings must be protected, eliminated completely or located out of predictable fire travel paths.

3. **Draft stopping, fire separations, and tenant separations need to be coordinated into a sound compartmentation system that will slow the rate of fire spread through unsprinklered combustible spaces.**

Post fire investigation suggested that the location of attic draft stop partitions was not as noted on the building's plans. Apparently the installed draft stops in the attic trusses did not coincide with the separation walls between apartments. Because the draft stops may have been located over apartments, as opposed to lining up with apartment separation walls, the compartmentation from the floor to the underside of the roof deck was not continuous. After the fire was established in an apartment a travel path around the draft stop and into the attic was provided.

Attic draft stops are intended to slow the rate of a fire's progression by providing a few minutes of delay. It is intended to provide a point where the fire department can take a defensive stand. A draft stop does not possess the same fire resistance as a fire partition or a fire wall, nor does it penetrate a roof, even when the roof is of combustible construction. It will not be an effective fire stop without fire department assistance and support.

4. **The Lodge Building had five separate fire department connections to provide fire department support to the automatic sprinkler system. Each wing had a separate connection and the fifth siamese was at the pit in front of the building.**

Multiple connections are difficult to deal with even when grouped together and clearly marked as to their function or area protected. In this instance, one siamese connection was highlighted by being prominently located at a pit in the front of the building and was used by one of the two first-in companies. There were no reports of any of the other four connections beings used. This could have resulted from the fire's location in the core area making access to the other connections dangerous. Alternately, the lack of their use could be the result of confusion on the number of connections needed to support the entire sprinkler system. Because of the extension of the fire into the unsprinklered attic space early in the incident, it is doubtful that the use of the four additional fire department connections would have had any material effect on the end result.

5. **Not all sides of the x-shaped structure were easily accessible for the fire department apparatus. Sector C contained the pool and recreation area which was surrounded by an iron fence and did not have direct vehicle access.**

The restricted access to the sides of the building made it difficult to directly attack the initial fire. It also reduced the effectiveness of the initial hose streams because they were unable to reach the entire fire area. Large caliber fire streams could not be rapidly placed onto all sides of the main fire body to wet uninvolved areas of the structure. Fire growth in Sector C constituted a problem that could not be completely resolved until more resources were available.

Access to all sides of a building should be incorporated into the design and construction process. Proposed site drawings should indicate the location of fencing, landscaping, walkways, and security features which may slow fire department access. Features, such as emergency access gates, wide sidewalks, fire lanes and reinforced all weather road beds, should be provided from the normal roadways into key locations.

6. **Calling for additional fire units early in the incident is important to support first alarm units when faced with multiple tactical operations.**

Tactical plans involving fires in large residential occupancies continually balance resources between search and rescue and fire suppression. Both must be done simultaneously because concentrating on one can result in neither being accomplished successfully. To complete search and rescue, the fire spread must be slowed to allow companies the time to work. All residents immediately endangered by the fire must be quickly evacuated. Even residents remote from the actual fire can be overcome from the smoke and the large quantity of carbon monoxide generated during fire suppression.

Additional fire companies were called prior to the arrival of the first due unit based on dispatch information and the fire officer's knowledge of this structure. After the arrival of the first in battalion chief, more fire units were dispatched to the incident. Additional units were dispatched to reinforce Sectors and provide relief crews for the Sectors.

Because of the large area, irregular shaped buildings, and the large volume of fire some support functions were decentralized to the Sector level. Staging was not established because arriving units were immediately assigned to Sectors to perform tactical operations or provide relief crews. Rehab areas were established on the Sector level to provide services to fire companies in a more effective and efficient manner.

7. **The extended detection time likely contributed to the difficulty in controlling the fire.**

Based upon the time of fire origin and its location on an outside walkway, it is believed that some delay occurred in detecting this fire. The nearest fire detection device to the fire's point of origin was the sidewall style automatic sprinkler opposite the open stairway. By the time this head operated and the waterflow alarm registered the activation, it is likely that the fire was already threatening the attic, if it was not already into the attic. The first fire department units on the scene reported a substantial fire in progress around the building's center core area.

8. **This structure made use of light-weight combustible surfaces and contained substantial unprotected combustible concealed spaces. These features provide for rapid fire spread and growth. Both officers and firefighters need to recognize and report on finding these items.**

This was the largest fire in a multi-family occupancy in the history of the Chesterfield Fire Department and it was successfully controlled without loss of life or significant injury to occupants or firefighters. Yet the fire did substantial destruction and damage to the property. The speed of fire travel and the time needed to establish an effective suppression operation are elements that must be incorporated into the overall tactical plan. These two elements influence the placement of individual companies, location of master streams, and the time search operations they will have available.

Individual and multiple company drills can identify the amount of time and the effort required for specific fireground operations. However, learning about and appreciating how fire spreads and how quickly it can expand is not easily done. This learning involves a combination of classroom theory, review of past fires, thorough post-incident critiques and actual experience. Each of these elements is different for firefighters, company officers, and incident commanders.

APPENDIX A

Old Buckingham Station

APPENDIX B

Lodge Building

APPENDIX C
Construction Details

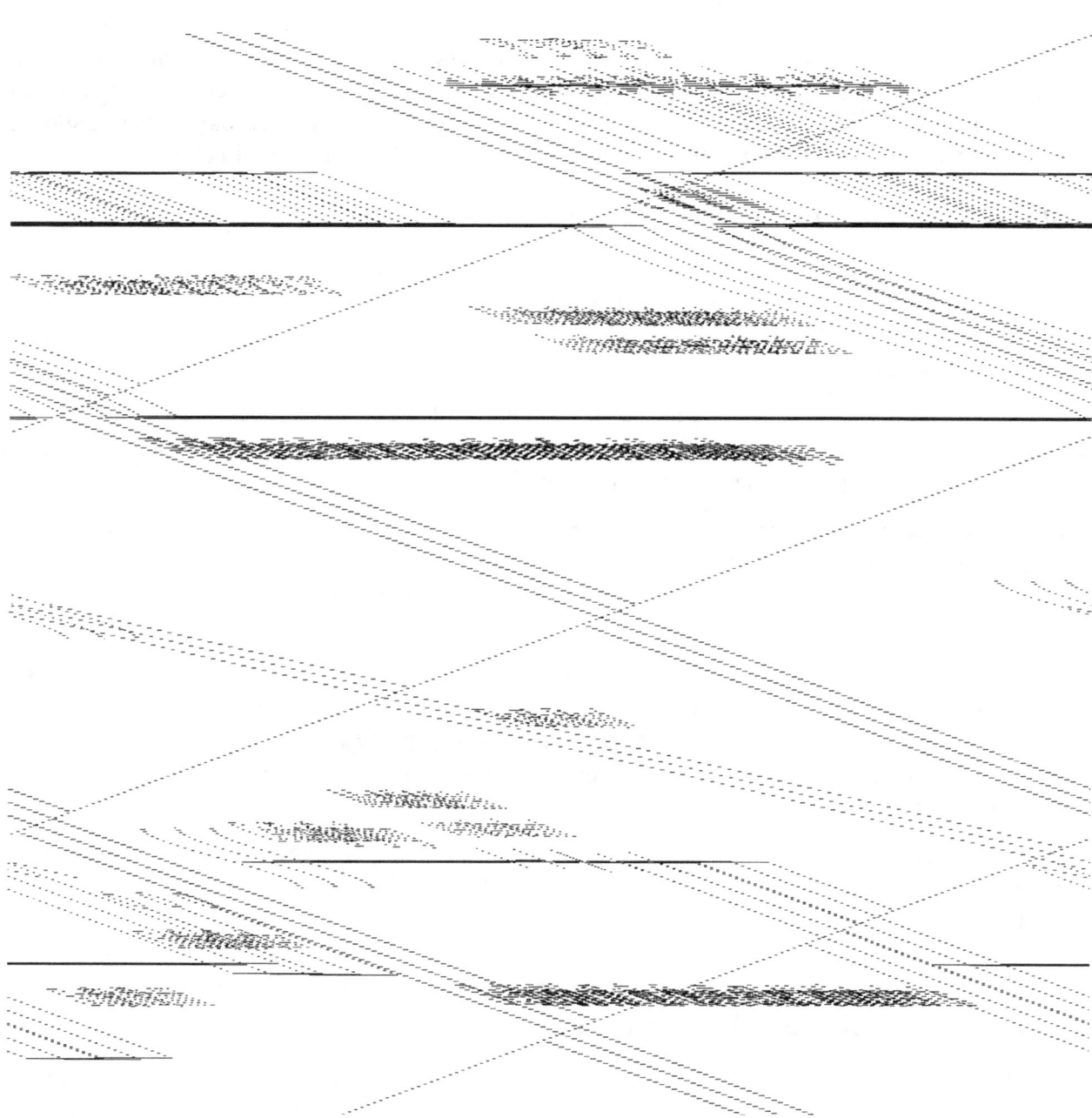

APPENDIX D

National Fire Protection Association Residential Automatic Sprinkler Requirements

This section contains a summary of residential sprinkler system design criteria for three NFPA automatic sprinkler system standards. Two of the standards were published and effective at the time that Old Buckingham Station was being designed and constructed. The third standard was being developed during this period and was officially published with an effective date of February 6, 1989.

NFPA Standard No. 13, 1985 edition, Installation of Sprinkler Systems

A new section in Chapter 7 on hydraulically designed systems provided the water supply requirements. This section applies to dwelling units where listed residential sprinklers are used. Other areas or the use of other types of automatic sprinkler heads would result in the application of other water supply requirements from the standard.

Water supply requirements specify a minimum of 18 gpm for a single operating sprinkler head and 13 gpm per sprinkler head for multiple operating sprinkler heads. This is the same as NFPA 13D because the same fire test data and occupancy is used.

All sprinkler heads in a compartment to a maximum of four heads shall be used in determining the total water supply amount. This represents two additional heads beyond that required by NFPA No. 13D. The typical water supply requirements would be about twice that for one-and two-family dwellings.

This standard requires that automatic sprinklers be provided throughout all parts of the building. This would include combustible walkways, breezeways, and balconies. For unheated spaces, either dry systems, antifreeze systems, or special dry style sprinkler heads would be required. In this building, automatic sprinklers would have been required in the attic, walkways and most likely the apartment balconies.

The presence of automatic sprinklers in the attic would have likely reduced the extent of destruction. However, NFPA Standard No. 13R applied to the sprinkler installation as the building was rebuilt and the attic was not protected.

NFPA Standard No. 13D, 1984 edition, Installation of Sprinkler Systems in One- and Two-Family Dwellings and Mobile Homes

The installed automatic sprinkler system would use only new, listed or approved residential style automatic sprinkler heads. The response and water distribution characteristics of other automatic sprinkler heads might require different water supplies.

Water supply requirements specify a minimum 18 gpm for a single operating sprinkler head and 13 gpm per sprinkler head for multiple operating sprinkler heads.

All sprinkler heads in a compartment to a maximum of two heads shall be used in determining the total water supply amount. Essentially, this means that the typical fire protection water supply would be at least 26 gpm for the dwelling. The pressure required would have to meet the requirements of the sprinkler head listing.

Automatic sprinklers may be omitted from typical bathroom and closets, open porches, garages, carports, and attics and crawl spaces which are not intended for living purposes or storage. The emphasis is to locate automatic sprinklers in the living area, heated utility areas, and unfinished storage areas.

NFPA Standard No. 13R, 1989 edition, Installation of Sprinkler Systems in Residential Occupancies up to Four Stories in Height

This is the first edition of the standard and represented a milestone in the design of automatic sprinkler systems for low rise multiple family dwellings. This standard was published after the occupancy certificates for this property were issued.

The water supply requirements are the same as NFPA Nos. 13 and 13D which specify 18 gpm for a single head and 13 gpm per head for multiple operating heads. Again, the same fire test data and occupancy are used for designing the protection.

All sprinkler heads in a compartment to a maximum of four heads are used in determining the total water supply amount. This is the same water supply as that required by NFPA No. 13.

Unlike NFPA No. 13, this standard does not require all spaces in the structure to be protected by automatic sprinklers. Bathrooms, small clothes closets, attics, crawl spaces, elevator shafts and exterior balconies, corridors and porches are areas where automatic sprinklers can be omitted.

APPENDIX E

Section at Core Walkway, Lodge Building

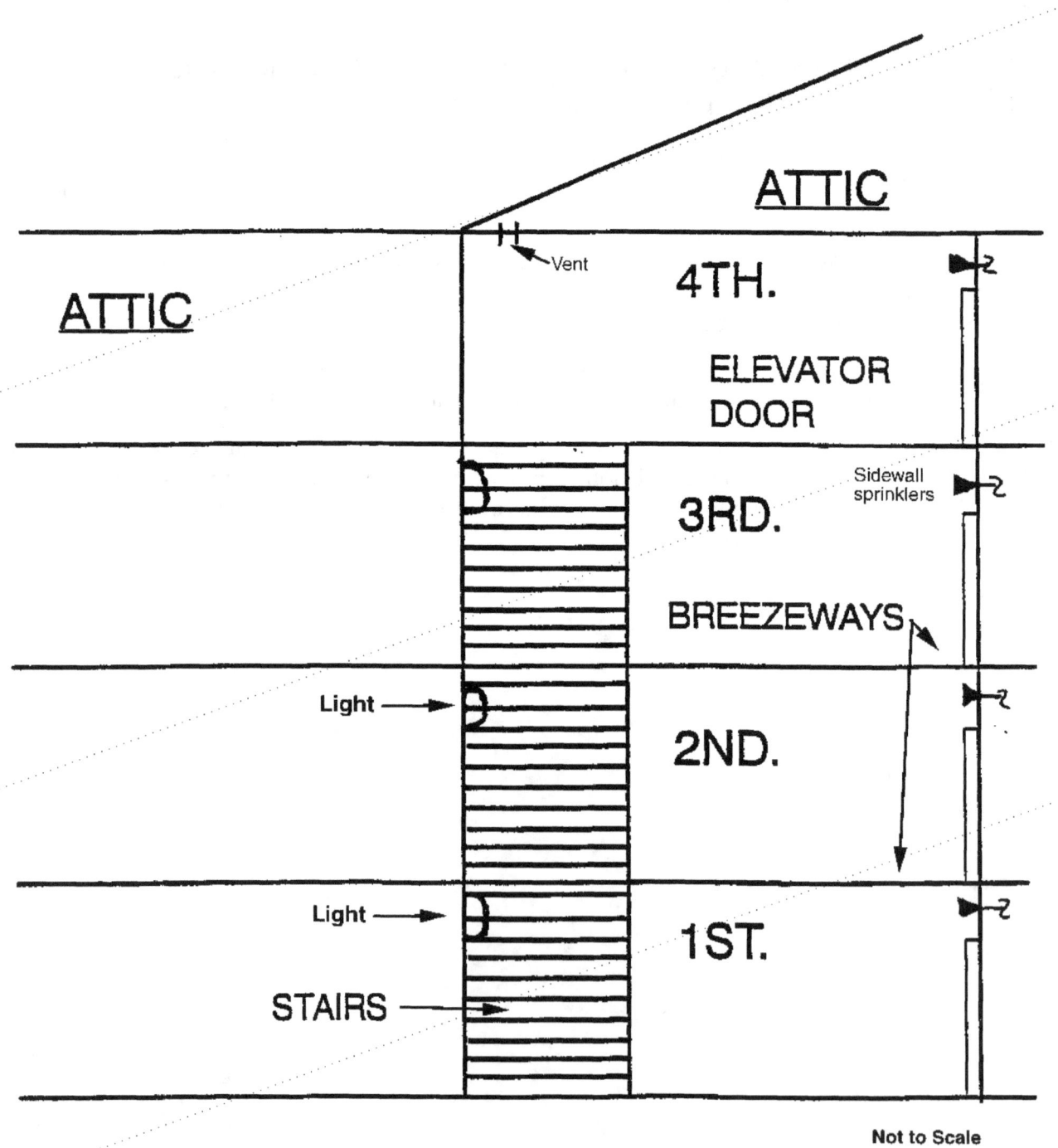

ATTIC

ATTIC

Vent

4TH.

ELEVATOR
DOOR

Sidewall
sprinklers

3RD.

BREEZEWAYS

Light

2ND.

Light

1ST.

STAIRS

Not to Scale

APPENDIX F

Tactical Operations, Initial

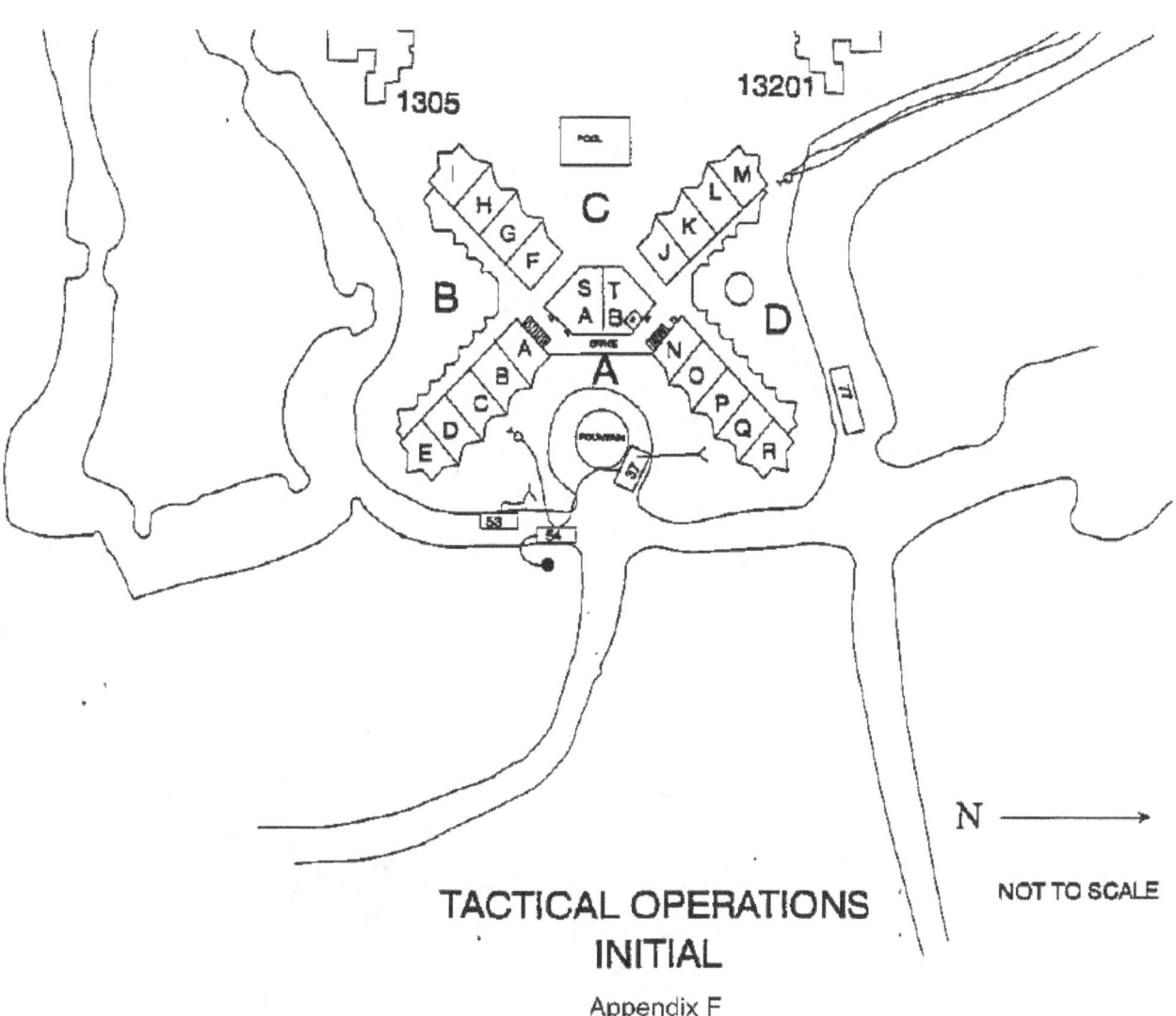

TACTICAL OPERATIONS
INITIAL

Appendix F

N ⟶

NOT TO SCALE

APPENDIX G

Tactical Operations, Search & Rescue

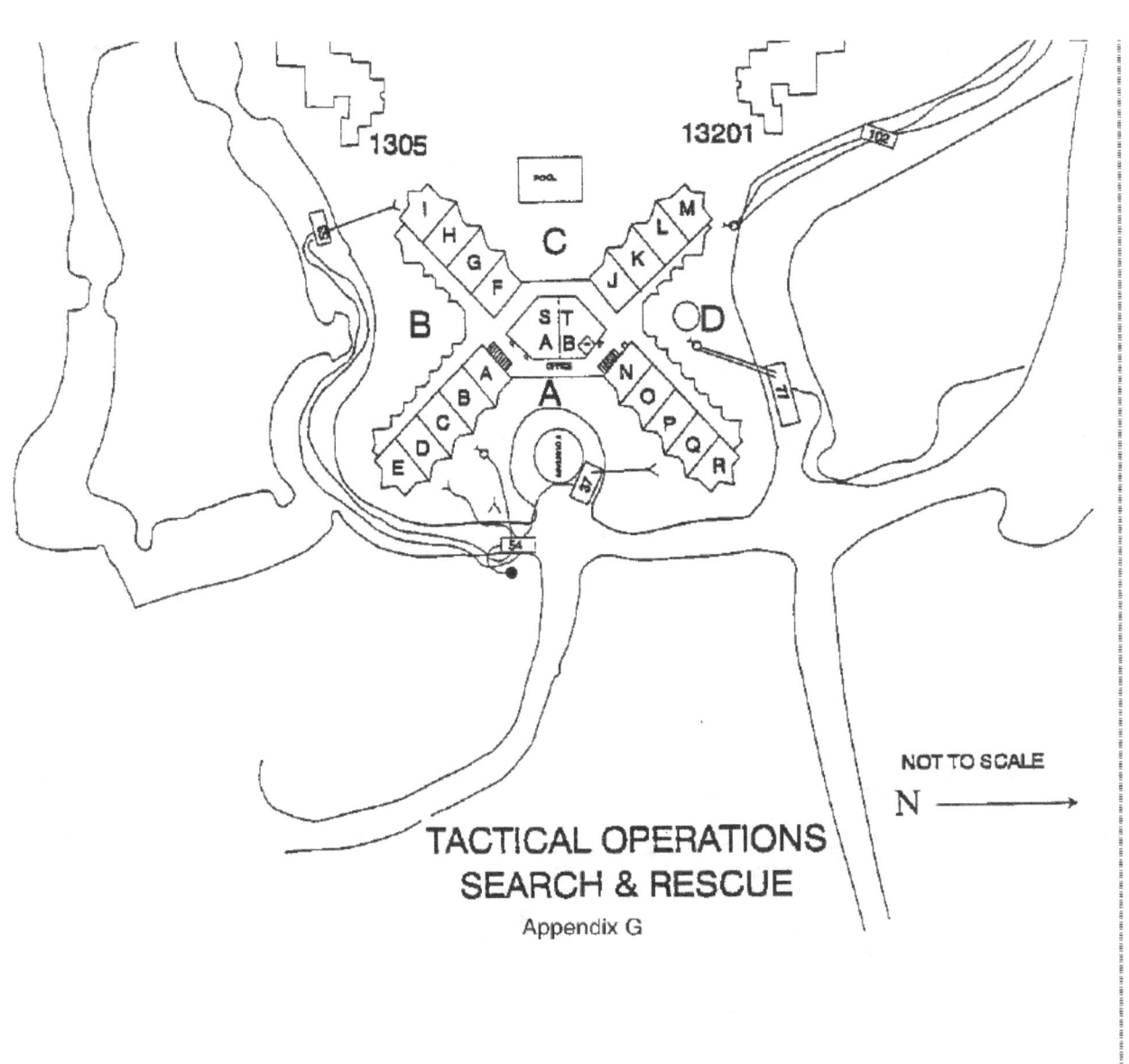

TACTICAL OPERATIONS
SEARCH & RESCUE
Appendix G

NOT TO SCALE

N ⟶

Tactical Operations, Suppression

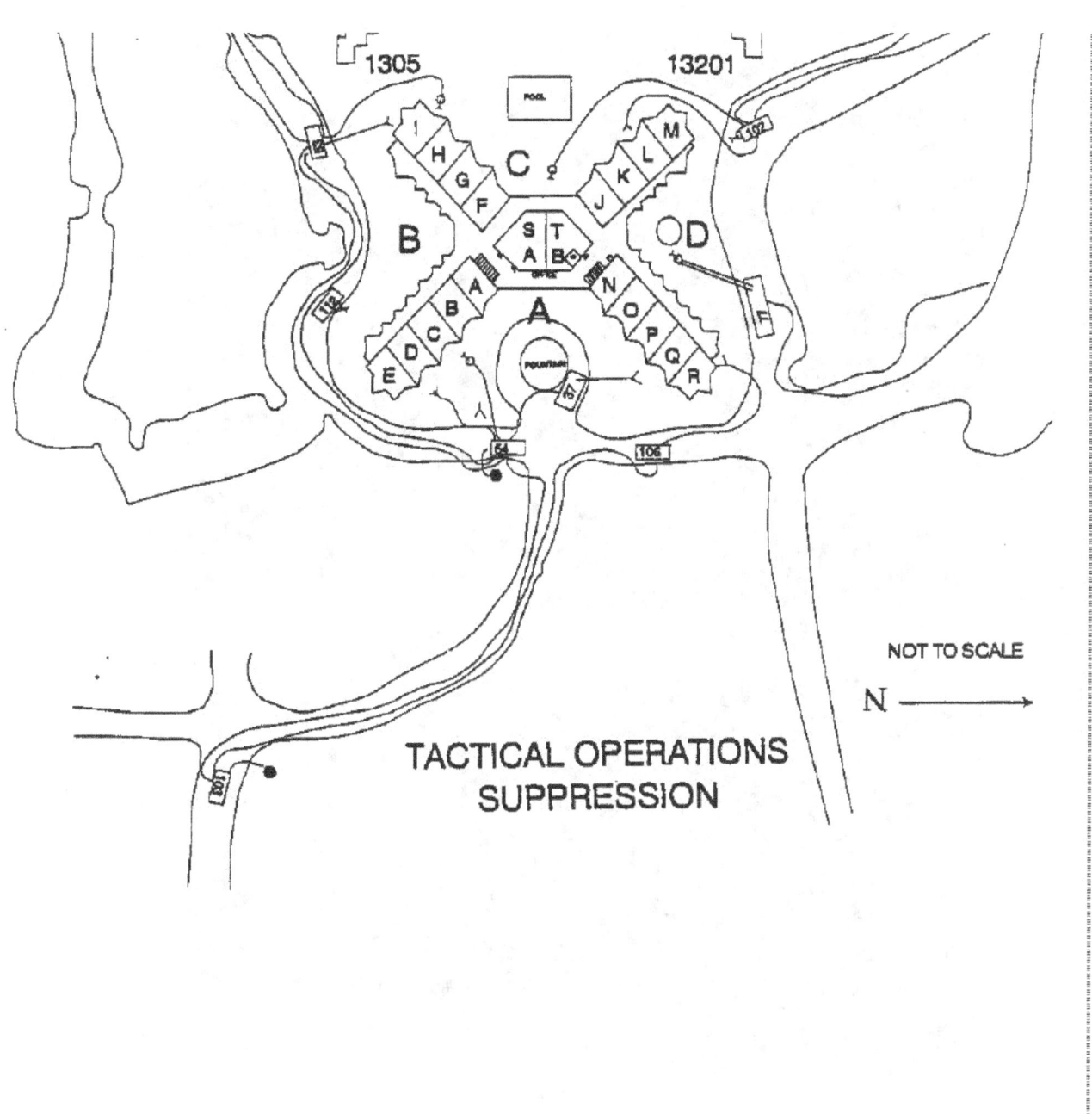

TACTICAL OPERATIONS
SUPPRESSION

NOT TO SCALE

N ⟶

APPENDIX I

Photographs

Photo 1. Site view of fire with north to the left. The shopping center and woods for the hand-stretched hose is on the left with Engine 73 still in position.

Appendix I (continued)

Photo 2. View of building with units still operating to complete extinguishment. North is to the left.

Appendix I (continued)

Photo 3. The main entrance to the building and the four story core building before the fire. North is to the right.

Photo 4. The rear or swimming pool side of the four story core building before the fire. The fire started on a second floor breezeway on the left side of this photograph.

Appendix I (continued)

Photo 5. Fire involvement in Sector D (North side) as units were preparing to operate in this sector. The fire was well-established in the attic, third floor walkway, and fourth floor.

Photo 6. Fire hydrant and core siamese connection used by Engine 54 on arrival. Repairs and reconstruction are being made to the front wing; this side is the patio/balcony for each apartment.

Appendix I (continued)

Photo 7. The remains of the front archway and the repairs and reconstruction underway for the wing in Photograph 6. The attic trusses and exterior sheathing materials are illustrated.

Photo 8. Reconstruction underway in Sector B, which is on the south side of the building. The exterior walkways were on the side of the wings.

Appendix I (continued)

Photo 9. The exterior walkways in sector D, with the stairway located at the left edge of the photograph. The core is out of the picture to the right.

Photo 10. The stairway and exterior walkways used throughout the building. Note the gaps in the walkway to allow water to drain and the piece of thin exterior sheathing leaning against the stairway.

Appendix I (continued)

Photo 11. Typical eave line vents installed around the perimeter of the building. The fire is believed to have entered the attic through these vents.

Photo 12. The parallel cord wood floor trusses used in the original construction and a polybutylene automatic sprinkler line.